Cheila Biazussi

Rhizogenesis of tung (Aleurites fordii) cuttings

Cheila Biazussi

Rhizogenesis of tung (Aleurites fordii) cuttings

Subjected to different doses of Indolebutyric Acid

ScienciaScripts

Imprint

Cover image: www.ingimage.com

This book is a translation from the original published under ISBN 978-613-9-70346-3.

Publisher:
Sciencia Scripts
is a trademark of
Dodo Books Indian Ocean Ltd. and OmniScriptum S.R.L publishing group

120 High Road, East Finchley, London, N2 9ED, United Kingdom
Str. Armeneasca 28/1, office 1, Chisinau MD-2012, Republic of Moldova, Europe
Printed at: see last page
ISBN: 978-620-8-18800-9

I dedicate this work, which symbolically represents six years of struggle and dedication in search of knowledge, professional and personal improvement, to all the people who in one way or another contributed to me reaching the end of this journey.

ACKNOWLEDGEMENTS

To God, the foundation of everything.

To the wise and benevolent spirits who support us in the trials of this life and are always present in a wise and great way, allowing the objectives of this life to be achieved and realised according to God's will.

To my mum, Vanda Biazussi, my source of light to walk all the paths, I thank you for all your love, support and understanding today and always.

To my grandmother Zenaide, my brothers Fabio and Cleiton, for whom I have a special affection.

I would also like to thank Ediméia and Davi for always being there for me in a joyful, pure and sincere way.

My special thanks go to Sabrina, Daniela, Andrieli and Juliana for their friendship, encouragement, support and inspiration.

To all my family who, in one way or another, contributed to this journey.

I would like to thank my friends and colleagues for their sincere friendship, joys, shared problems and mutual support, which has made us better and stronger people over the years.

To the Agronomy course and all the teachers who contributed to my personal and professional development.

My special thanks go to Professor Dr Margarida Flores Roza Gomes for her patience, support and guidance in the construction of this work.

I would also like to thank professor and coordinator Dr Alceu Cericato for his support and encouragement, and for sparing no effort to ensure that Agronomy students cross the line of comfort in search of new experiences and knowledge.

Study yourself, noting that self-knowledge brings humility and without humility it is impossible to be happy.

(Allan Kardec)

SUMMARY

The tung (A/eurites *fordii* Hemsl) stands out for the high concentration of oil in the fruit (around 45 per cent). Tung oil stands out on the market for its high drying power, resistance and chemical stability, which is why it was widely used in China for preserving ship hulls, in the USA for preserving military equipment and currently by the furniture market. Tung needs 300 to 400 hours of cold to break dormancy and is currently grown in Rio Grande do Sul, a state with a climate similar to the far west of Santa Catarina. This has led to the need to assess the best way of propagating tung to suit the region's climatic conditions. The aim of this study was to evaluate the performance of woody, semi-woody and herbaceous tung cuttings submitted to different doses of indolbutyric acid. The experiment was carried out in a greenhouse located in the experimental area of the University of Western Santa Catarina (UNOESC), situated in the town of Maravilha (SC). The experimental design was entirely randomised (DIC). In a 3x3 factorial design, with three doses (0%, 50% and 100% of indolbutyric acid (AIB)) and three types of cuttings (woody, semi-woody and herbaceous) with three replications, the plots consisted of four plants, making 36 cuttings per treatment, totalling 108 cuttings. The data was subjected to Analysis of Variance (ANOVA) using the F test and the means were compared using the Tukey test at 0.05 significance level. The tung cuttings where 100% of the recommended dose of AIB was used showed positive results and the dose favoured the development of the tung cuttings, which showed a greater occurrence of callus and better initial growth.

Keywords: Adaptability. *A/eurites fordii* Hemsl. Rooting. Western Santa Catarina.

SUMMARY

CHAPTER 1

INTRODUCTION

The tung (A/eurites *fordii* Hemsl) is a plant from the Euphorbiaceae family, native to Asia, predominantly in China and currently produced on a large scale in South America, the United States and Africa (OLIVEIRA; GODOY; COSTA, 2003). According to studies carried out by Silva et al. (2013), the tung bean has a high oil production potential, with an average kernel content of 47%, which can reach 63% of its weight, which shows an average production of 1,327 kilos of oil per hectare, which when transformed into biodiesel has a yield of 87%, according to (KAUTZ et al., 2008).

As a species native to temperate climate regions, it requires between 350-400 hours of cold at temperatures below 7.2 °C in order to break dormancy and stimulate flowering and budding in early spring (ALMEIDA; SILVA; WREGE, 2012).

Data from IBGE (2012) show that commercial tung production in Brazil is concentrated only in Rio Grande do Sul, occupying around 400 hectares of cultivated area, mainly in the municipalities of Veranópolis - RS, Fagundes Varela - RS and Cotiporã - RS, a fact that is also related to the scarcity of information on this crop. According to Duke (1983), research from the western United States was used as a basis for establishing the crop in Brazil, but research in Brazil began to receive greater encouragement after the launch of the biodiesel programmes, created in 2004, as an alternative market for soya.

According to Almeida, Silva and Wrege (2012), because it is a perennial tree-like plant with a deep root system, tung can tolerate short periods of drought without deeply compromising fruit production. The most serious climatic risk for this crop is the occurrence of late frosts that can hit the most fragile tissues of the flowers and fruit in formation, causing a drop in production.

The lack of information on this plant and the installation of oil processing plants only in the Serra Gaúcha contributes to limiting production. Therefore, there is a need to carry out studies on the adaptation of this plant in other regions of the country, spreading production in order to provide alternative income for farmers living on less fertile land with uneven soils, which cannot be used for annual crops or fruit trees, but which can be adapted to the production of tung, which has an industrialisation and processing market that is already consolidated, although it is not very widespread.

The aim of this study is to evaluate the performance of woody, semi-woody and herbaceous cuttings of tung (A/eurites *fordii* Hemsl) submitted to different doses of indolbutyric acid, in order to produce technical and scientific information related to the adaptation of tung in the western region of Santa Catarina, as well as the best way to grow and propagate it.

1.1 PRESENTATION OF THE THEME

Woody, semi-woody and herbaceous cuttings of tung (A/eurites *fordii* Hemsl) subjected to different doses of indolbutyric acid.

1.2 PROBLEM

Considering the adaptation and cultivation of tung in other regions of Brazil with similar climatic conditions and terrain to the west of Santa Catarina, it is believed that the adaptation and acceptance of tung should be studied as an alternative way of supplementing income, given that the predominant economic activity in the region is dairy farming, which has seen considerable market fluctuations.

It can be seen that small properties, a cultural characteristic of western Santa Catarina, most of which are located on hilly terrain, need an alternative source of income for their less productive areas.

Tungue adapts to regions with uneven soils and low nutrient levels. The limiting factor is propagation, which, according to Silva et al. (2013) is expensive through micropropagation and loses important genetic characteristics when done by sowing.

Micropropagation is a technique for developing shoots in a controlled environment, usually in the laboratory (in *vitro). It* is carried out from the buds emitted by the plant and induces the formation of new buds. Micropropagation makes it possible to obtain more seedlings in a shorter period of time when compared to other seedling propagation methods (ALVES, 2015).

The process of cuttings is the most suitable for tung propagation, as it allows the propagation of plants with good genetic characteristics. This process is strongly influenced by the type of cuttings, their age, the physiological condition of the parent plant and the time of collection, factors which will be specified in this study.

To help the adaptation and development of tung in the west of Santa Catarina, doses of indolbutyric acid will be evaluated. This is an easy-to-acquire and low-cost growth regulator, which facilitates acceptance and does not increase production costs. Thus, the indications that motivate this study are:

a) What are the possibilities for adapting tung in the far west of Santa Catarina?

b) What dose of growth regulator will give the most efficient results in the development of the cuttings?

1.3 HYPOTHESIS

- HO: Indolebutyric acid does not significantly interfere with the initial growth of tung (A/errifes *fordii* Hemsl) cuttings.

- H1: Herbaceous cuttings of tung *(Aleurites fordii* Hemsl) have a higher percentage of cuttings with leaves than woody and semi-woody cuttings.

1.4 BACKGROUND

Considered a noble crop with few studies and of great economic importance in other regions of southern Brazil. Tung develops in less fertile regions and does not require flat terrain. For these reasons, combined with the economic feasibility research carried out previously, it is believed that adapting tung in the Far West region of Santa Catarina could be an alternative source of income for the local community.

Given the existence of studies on the economic viability of the activity in the region and the lack of studies carried out in the field, there is a lack of concrete information on the adaptation of the crop to the regional microclimate.

This study aims to identify the best way to propagate this crop, considering that its agronomic characteristics are in line with the reality of the region and its economic potential could surpass existing ones.

1.5 OBJECTIVES

1.5.1 **General Objective**

To evaluate the performance of woody, semi-woody and herbaceous cuttings of tung (A/eurifes *fordii* Hemsl) subjected to different doses of indolbutyric acid.

1.5.2 **Specific objectives**

- Determine the percentage of live cuttings;
- Quantify the cuttings with callus;
- Identify the most appropriate type of stake for propagating tung;
- Specifying the most appropriate dose of growth regulator for the development of tungsten cuttings.

CHAPTER 2

THEORETICAL FOUNDATIONS

2.1 TUNGUE

(A/eurifes *fordii* Hemsl) is a perennial plant that begins its production cycle in the third year, with active commercial production from the fifth year onwards, remaining active for up to 30 years. In addition, it is estimated that it shows good adaptation indicators in western Santa Catarina. To this end, practical applications and studies are needed on its resistance to the adversities of the regional microclimate, as well as conclusions on the most suitable hormonal treatments for its full development (UCZAI, 2009).

2.1.1 **History**

The tung *(Aleurites fordii* Hemsl) grown in large quantities in China has its origins in Asia, in the so-called Far East at 30° North latitude, where it was commonly used in the manufacture of ships, more specifically applied to preserve the hull because it stands out for its hardness and resistance to humidity. Tungue later spread to Africa, South America and the United States in 1905 and from 1927 onwards it was cultivated in Florida, Georgia, Alabama and Mississippi (NETO, 2008).

It has been used since the 16th century as a protective oil for wood, and is only rated lower than linseed oil in the manufacture of varnishes, as it has a higher drying index (BARROS, 2016).

The Chinese used tung oil as a waterproofing agent for ship hulls, masts and sails. The USA used it to coat parts in the Second World War because it was resistant to base acids (BUTTERFLY OIL, 2017).

In Brazil, oil extraction was first recorded in Florianópolis by a rudimentary plant in 1925. In São Paulo, seedlings were introduced in 1928 from Florida (USA) by the director of the Agricultural School of Viçosa, Minas Gerais. In 1930, cultivation began in Tatuapé (SP) through Indústrias Reunidas Fabricas Matarazzo, which bought seedlings from Georgia (USA) (NETO, 2008).

The tung was processed in a rudimentary way using a centrifugal peeler, which threw the nut against a plate inside a compartment. After separating the nuts from their shells, the kernels were ground in a common meat grinder. The mass of strands was pressed into sacks to extract the oil (NETO, 2008).

2.1.2 **Agronomic characteristics**

The tung (A/eurifes *fordii* Hemsl) is a plant grown in poorly fertile soils, preferably stony, with a pH between 6.0 and 6.5, tolerating between 5.4 and 7.1. It can reach a height of 9 metres and requires a period of winter dormancy, for which it needs 350 to 400 hours of cold (<7.2°C) (ALMEIDA; SILVA; WREGE, 2012).

The development of tung is favourable in regions with long, hot summers and high rainfall (1,120 mm of annual rain), flowering in September and October (growth branches from the previous season) (REITZ, 1988).

With regard to its climate vulnerability, tung can show a drop in production when exposed to late frosts that hit the most fragile tissues of the flowers and fruit that are just beginning to form (GOLFETTO et aL, 2011).

2.1.3 **Management**

Ávila (2010) states that the propagation of tung by seed presents low standardisation in the type of plants, causing difficulty in cultural treatments such as pruning, spacing, fertilisation, control of invasive plants and uniformity in the harvest. Some seed-based commercial plantations have spacings of 6 to 8 metres between rows and 4 to 8 metres in the rows, but no pruning or other cultural treatments are carried out.

According to data from Epagri (2001), the average lowest winter temperatures in the far west of Santa Catarina are between 7 and 9 °C (Figure 1), which would meet the crop's needs.

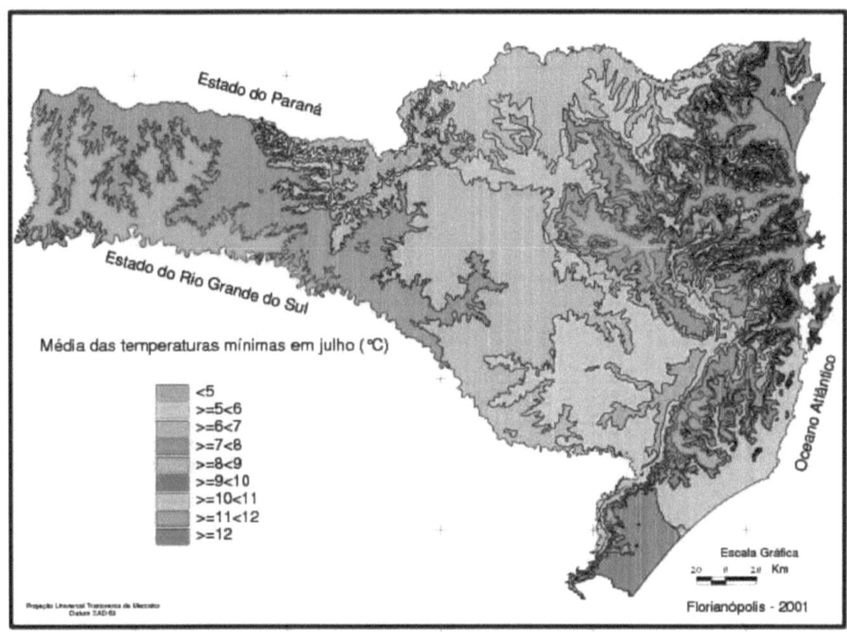

Figure 1- Climatological Atlas of the State of Santa Catarina
Source: Epagri, (2001).

Bertocchi (2011) carried out quantitative studies comparing the net results of tung cultivation, considering as an initial investment the land for tung cultivation and the pre-planting investment for corn planting, in an area of one hectare. The research results were based on data presented by the National Supply Company (CONAB) and the Santa Catarina Institute of Agricultural Planning and Economics (ICEPA). The author concludes that tung is more profitable than maize, one of the main economic activities in the region, as shown in Table 1.

Table 1- Comparison of yields between tung and maize crops in the Far West of Santa Catarina

PRODUCT	YEAR 1	YEAR 2	YEAR 3	YEAR 4	YEAR 5	YEAR 6	YEAR 7	YEAR 8	YEAR 9	YEAR 10
TUNGUE	17.670,00	1.531,67	1.531,67	468,33	2.468,33	3 598,33	5.598,33	7.596,33	9.598,33	11.358,33
CORN	-9512,20	487,80	487,80	487,80	487,80	487,80	487,80	487,80	487,80	487,80
DIFFERENCE	-8157,80	-2019,47	-2019,47	-19,47	1980,53	3110,53	5110,53	7110,53	9110,53	10870,53
BALANCE ACCUMULATED	-8157,80	-10177,27	-12196,73	-12216,20	-10235,67	-7125,13	-2014,60	5095,93	14206,47	25077,00

PRODUCT	YEAR 11	YEAR 12	YEAR 13	YEAR 14	YEAR 15	YEAR 16	YEAR 17	YEAR 18	YEAR 19	YEAR 20
TUNGUE	11.35833	11.358,33	11.358,33	11.358,33	11.358,33	11.358,33	11 358,33	11.358,33	11 358,33	11.358,33
CORN	487.80	487,80	487.80	487,80	487.80	487,80	487,80	487,80	487.80	487,80
DIFFERENCE	11846,13	10870,53	10870.53	10870,53	10870,53	10870,53	10870,53	10870,53	10870,53	10870,53
BALANCE ACUMULATED Q	36923,13	47793,67	58664.20	69534,73	80405,27	91275,80	102146,33	113016,87	123887,40	134757,93

PRODUCT	YEAR 21	YEAR 22	YEAR 23	YEAR 24	YEAR 25	YEAR 26	YEAR 27	YEAR 28	YEAR 29	YEAR 30
TUNGUE	11.358,33	11358,33	11.358,33	11.358,33	11.358,33	11.358,33	11.358,33	11.358,33	11.358,33	11.358,33
CORN	487.80	487,80	487.80	487,80	487,80	487,80	487,80	487,80	487,80	487,80
DIFFERENCE	11846,13	10870,53	10870,53	10870,53	10870,53	10870,53	10870,53	10870,53	10870,53	10870,53
BALANCE ACCUMULATED O	146604,07	157474,60	168345,13	179215,67	190086,20	200956,73	211827,27	222697,80	233568,33	244438,87

Source: Bertocchi (2011).

The oil is extracted from the fruit, but it must be harvested by hand when the fruit falls to the ground, as it is not uniformly ripe. Harvesting should begin when four or more fruits begin to fall and the humidity is around 30 per cent, but for delivery to the industry the ideal humidity is 14 per cent (REITZ, 1988).

The average yield of tung per hectare is 34.4 tonnes (SILVEIRA et al.,

2010), but in areas where farmers choose not to invest in fertiliser, the average yield is 18 tonnes of dried fruit per hectare. After peeling, the seeds account for 45 per cent of the total (BERTOCCHI, 2011).

2.1. 4**Chemical** properties

Tung is a fixed oil, i.e. it has a high molecular weight, which gives the product a high oil yield of approximately 33 per cent (GOLFETTO et al., 2011).

With a density, tested at 25°C, of 0.920 - 0.938 g/cm^3 , the saponification index of tung oil is 189 - 198mgKOH/g, and the iodine index can vary between 158 and 178 Cg 12/g. The oil also has a refractive index at 25° C of 1.5160 - 1.5260 and the gelatinisation time at 280°C varies between 10 and 13 minutes (BARROS, 2016).

Research by Golfetto et al. (2011) shows that the quick-drying quality of tung oil is attributed to the presence of oleostearic acid. Its impermeability and resistance to chemicals is due to the natural tendency of molecules to bond, reinforced by the cooking process which alters the structure of the molecules (BUTTERFLY OIL, 2017).

The ideal packaging for tung oil is in metal or plastic drums, which gives the oil an approximate shelf life of 24 months (BARROS, 2016).

2.1.5 **Current market**

Tung is characterised by its high oil content, which, when extracted from the seed, becomes the main product obtained from this plant and is used in the chemical, paint (varnish and resin), printing and wood industries. The main difference between tung oil and other oilseeds is its specific quick-drying properties, which ensure that it can be used in the paint industry, as well as for sealing wood (ALMEIDA; SILVA; WREGE, 2012).

Gomes et al. (2008) presented studies related to the efficiency of tung as a nematicide. Pretto (2013) carried out tests using tung bran in fish feed, analysing the composition of the bran and its digestibility, obtaining positive results for both, with the averages for crude protein content being 27%, 3.3% oil and 52% NDF (Neutral Detergent Fibre).

According to experiments carried out by Golfetto et al. (2011), tung is within the acceptable standards of the ANP (National Petroleum Agency) for use as biodiesel and can be produced in a viable and sustainable way in the south of the country.

2.1.5.1 Application

Currently, tung oil stands out for its purity, resistance to humidity and colour stability, as it doesn't darken over time like other finishes and resists scratches, acids and alcohol. As it is a penetrating oil, it protects the porous substrates of wood and natural stone, and is applied to concrete, marble and metal to protect against rust (GENERAL IRON FITTINGS, 2017).

Tungue is wrapped in a very hard bark that can be used as a substrate for flower farms and as a source of energy for burning in boilers. Tungue oil is extracted from the seeds (on average 4 to 5 seeds) inside the fruit (EMBRAPA, 2008).

Another specificity of tung oil is that it quickly saturates wood fibres, does not mould, does not exude and forms an inert, non-reactive layer. As such, its application is classified as safe and suitable for surfaces in contact with food (FDA Regulation Title 21 CFR175.300), toys (European Standard EN-71) and indoor and outdoor furniture (GENERAL IRON FITTINGS, 2017).

The temperature of the surface to be applied tung oil must be between 12.5 °C and 24 °C and the humidity must be less than 65 % (BUTTERFLY OIL, 2017). The oil yield is 250 ml per 3 m^2 and the interval between applications must be 24 hours so that the oil is properly absorbed by the wood. After the first application, the surface should be left to rest for 40 minutes. Tung oil is currently sold for an average of R$35.00 per litre (GENERAL IRON FITTINGS, 2017).

2.2 PLANT HORMONES

Plant hormones are naturally occurring organic compounds produced in the plant. When they occur in low concentrations, they are capable of promoting, inhibiting or modifying morphological and physiological processes in the plant. Like growth regulators, synthesised substances also cause similar reactions in plants (CASTRO; VIEIRA, 2001).

Growth regulators are synthetic substances which, when applied to plants, produce effects similar to those of hormones, but must be in the concentrations required for each type of tissue and species (DALL'ORTO, 2011).

According to Melo (2003), the treatment of cuttings with growth regulators aims to accelerate the process of root formation in each cutting, increase root emission and uniform rooting, all of which are fundamental factors for successful production.

Due to the identification of a large number of natural hormones, as well as the production of new synthetic substances with similar functions, the name adopted by Dutch botanist Frits W. Went (1957) was adopted, where the term "hormone" is reserved only for natural compounds and the term "regulators" for synthetic compounds. The main plant hormones are Auxins, Gibberellins, Cytokinins, Abscisic Acid and Ethylene (FLOSS, 2011).

2.2.1 Auxins

According to Zeiger (2006), plants have various types of auxin. These auxins perform various functions in plant physiology, such as stem and root growth, which occurs through the process of plant cell elongation. When the plant has a low amount of auxins, its roots can grow, but the stem does not develop. On the other hand, a high concentration of auxins can cause the stem to grow, leaving the roots underdeveloped.

The auxins produced by the apical meristem of the stem reduce the activity of the axillary buds near the apex, and when the apical bud is removed from the plant, lateral branches, leaves and flowers appear. In addition to Tropisms, movements related to plant growth according to natural stimuli, such as phototropism, which is the movement of plants in reaction to light stimuli (RAMO, 2012).

According to Floss (2011), auxins promote the growth of stems, leaves and roots. They are synthesised in the apical meristems, auxiliary buds, young leaves and other regions of active growth. Auxin synthesis is greater in the light than in the dark, and in perennial plants in temperate climates auxin levels vary according to the seasons, with the highest concentrations in spring and summer and the lowest concentrations in autumn and winter (dormancy and rest).

The recommended dosage of auxins is below 0.5 mg/L, with the exception of AIA (indole-3-acetic acid) which can be used in higher concentrations as it remains less stable in the culture medium (RAMO, 2012).

Some synthetic substances are used in extremely small concentrations (ppm - particles per million) with an auxin function, such as naphthaleneacetic acid (ANA), dicamba (2-methoxy-3,6-dichlorobenzoic acid), 2-4-D (2,4-dichlorophenoxyacetic acid) and 2,4,5-T (2,4,5-trichlorophenoxyacetic acid). However, when used in higher concentrations, these constitute the group of hormone-selective herbicides used to control weeds in poaceae (cereals) (FLOSS, 2011).

2.3 INDOLBUTYRIC ACID

Indolbutyric acid (IBA) is characterised as an auxin that acts to stimulate rooting due to its greater chemical stability in the plant and lower photosensitivity and mobility (BASTOS et al., 2009).

According to Gontijo et al. (2003), the use of growth hormones in the propagation of cuttings by cuttings is of great importance to overcome the difficulty in rooting the cuttings, with auxin being used most frequently, as it provides considerable results in the rooting of cuttings, as it stimulates the synthesis of ethylene, favouring the emission of roots.

In the method of propagation by cuttings, the use of growth regulators, especially AIB, which is considered a synthetic auxin, plays a significant role in the emission of roots. It is mainly used in species with rooting difficulties and can be used in various species to accelerate root formation (VERNIER; CARDOSO, 2013).

According to Dall'Orto (2011), the use of growth regulators to stimulate cell growth and multiplication is of fundamental importance for the success of cuttings.

As some plants have enough hormone for root initiation and others don't, synthetic growth regulators are an alternative to make the production system more efficient (MELO, 2003).

2.4 VEGETATIVE PROPAGATION

Vegetative propagation is possible due to the capacity to store genetic information and the regenerative power of certain plant organs such as stems, roots or leaves, which, through this information and favourable environmental conditions (totipotency), are capable of regenerating and developing a new plant (ONO; RODRIGUES, 1996). Vegetative propagation does not involve meiosis or fertilisation, so identical reproduction of a plant's genotype is possible (XAVIER, 2002).

According to Wendling (2003) the use of vegetative propagation makes it possible to produce individuals that are identical to the mother plant, ensuring their genetic characteristics, facilitating the selection of plants with the highest economic return and making genetic improvement programmes more effective, as vegetative propagation makes it possible to form clonal plantations with high productivity and uniformity, as well as greater resistance to pests and diseases and the quality of products and by-products. However, the frequent use of vegetative propagation can have some disadvantages, such as narrowing the genetic base of clonal plantations.

Among the most common forms of vegetative propagation is micropropagation, which consists of using parts of the plant such as callus, organs, cells and protoplasts in tissue culture in test tubes with nutrient solutions and hormones (SILVA, 2005).

The macropropagation method consists of using cuttings and grafts, where grafting involves inserting the upper part of a plant, called the epibiotic, into another called the hypobiotic. Cuttings selected from clonal mini-gardens are rooted, and the rooting process is directly influenced by the quality and age of the genetic material (SILVA, 2005).

The results of an experiment using vegetative propagation vary according to the physiological conditions of the mother plant, maturity, size, type and time of collection of the propagule, as well as the mineral nutrition of the mother plant, growth regulators, light, temperature, humidity and propagation technique. Currently, cuttings are the most widely used method of vegetative propagation on a commercial scale (WENDLING, 2003).

2.5 STAKES

Cuttings can be defined as the reconstitution of a plant whose main advantage is the complete fixation of all the characteristics of its ancestor, transforming the vegetative part detached from the mother plant into a plant that is genetically identical to it (PRATI et al., 1999).

According to Tofanelli, Rodrigues and Ono (2003), using the cuttings method, it is possible to observe that some species have an easy time emitting adventitious roots from their cuttings, while others have difficulty rooting.

According to Prati et al. (1999), vegetative propagation by cuttings is an advantageous alternative because this method provides early production and preserves the characteristics of the original variety.

In the process of cuttings, stem cuttings are differentiated by the degree of lignification and are classified as herbaceous, semi-woody or woody. Herbaceous cuttings are prepared from herbaceous stems located at the ends of the branch. Semi-woody cuttings are intermediate cuttings located in the middle of the branch and woody cuttings are primary cuttings, those located closer to the stem (CUNHA, 2009).

2.6 RIZOGENESIS

The roots have the main function in the plant of storing and transporting the inorganic substances absorbed, as well as producing certain plant hormones. During their growth, roots occupy spaces left by older, dead and rotten roots, following a path of least resistance (USP, 2002).

The plant's first root, which originates in the embryo, is called the primary root and can be axial or pivotal, both of which give rise to secondary roots, also known as lateral roots. As the plant grows, there is a need for a balance between its total surface area that manufactures food (photosynthesising) and its total surface area that absorbs water and mineral salts (USP, 2002).

According to Cunha (2009), adventitious rooting is influenced by various nutrients that are important at each stage of the process, so mineral nutrition is considered a determining factor in rooting.

CHAPTER 3

MATERIAL AND METHODS

The experimental design used in this study was propagation by cuttings, the most suitable method currently used in tung cultivation. The methodology recommended by Embrapa (MELO, 2003) was used to collect, analyse and interpret the data.

3.1 DELIMITATION OF THE EXPERIMENT

The experiment was carried out in the greenhouse located in the experimental area of the Universidade do Oeste de Santa Catarina (UNOESC), Maravilha campus, situated in the municipality of Maravilha, Santa Catarina (Figure 2). The geographical coordinates of the experimental area are 26° 45' 55" S/ 53° 11' 43" W and an altitude of 574 m (GOOGLE EARTH, 2017). The study was conducted in a greenhouse between 03 September and 14 December 2015, consisting of a period of 105 days.

Figure 2 - Location of the experimental area
Source: Google Earth (2017).

3. 2 EXPERIMENTAL DESIGN

According to Campos (2000), in order to obtain a statistical analysis of the data collected in a field study, it is necessary to outline the entire research plan, defining the variables, the variation factors and the number of repetitions. Factorial experiments are experiments involving two or more factors, with the main effect being a change in the response variable due to a change in the level of the factor. Their advantage is that they provide more comprehensive results, allowing the effects of interactions between factors to be studied by estimating the effects of one factor within the levels of the other factor.

The experimental design of a research study can be defined as the way in which the treatments (levels of a factor or combinations of factor levels) are assigned to the experimental units (MALHEIROS; PANOSSO, 2015).

The experimental design for this study was entirely randomised. The factorial scheme was 3x3, with three doses (control, 50% and 100%

indolbutyric acid) and three types of cuttings (woody, semi-woody and herbaceous) with three replications and plots made up of four plants, totalling 36 cuttings per treatment, giving a total of 108 cuttings. These were:

H0: Untreated herbaceous cuttings (control or control).

S0: Semi-woody cuttings without treatment (control or control).

L0: Woody cuttings without treatment (control or control).

H50: Herbaceous cuttings with 50% indolbutyric acid.

S50: Semi-woody cuttings with 50% indolbutyric acid.

L50: Woody cuttings with 50% indolbutyric acid.

H100: Herbaceous cuttings with 100 per cent indolbutyric acid.

S100: Semi-woody cuttings with 100 per cent indolbutyric acid.

L100: Woody cuttings with 100 per cent indolbutyric acid.

3.3 DATA COLLECTION TECHNIQUES

The cuttings procedure was carried out in recycled 1 litre tetra pak boxes, where a 1:1 ratio of pine bark-based substrate and

fine-grained expanded vermiculite, with a water retention capacity (WRC) of 50 per cent and a pH of 6.5.

The cuttings were cut transversely in order to expose the tissues and increase the interaction between the applied phytohormone and the plant's nutrient translocation system. The cuttings were then dipped for five seconds in an aqueous solution containing indolbutyric acid and then placed at a depth of 4cm.

AIB (indolbutyric acid) is a hormone used to stimulate root growth, the recommended dose of which is 5mg/L of water (RAMO, 2012).

The experimental units were packed and kept in a greenhouse sharing the same temperature, humidity and light.

Data collection was carried out according to the methodology used by Embrapa (2013) for tung cuttings, where the transplanted cuttings remained in the tetra pak boxes for 105 days until they were removed for evaluation. The evaluation of the percentage of live cuttings, as well as the identification of the number of cuttings with leaves and cuttings with calluses was done manually.

To determine the number of live cuttings, cuttings with leaves or green shoots were identified, counting the total number of cuttings per treatment.

To identify the cuttings that had developed the aerial part, the presence of leaves was analysed, obtaining the total average per plot and then per treatment.

To analyse the percentage of cuttings with callus, the behaviour of the cuttings was identified by obtaining the total average per plot and then per treatment.

3.4 DATA ANALYSIS AND INTERPRETATION TECHNIQUES

The data was subjected to Analysis of Variance (ANOVA) using the F test and the means were compared using the Tukey test ($P<0.05$). The software used was Sisvar 5.6 (Build 86) (FERREIRA, 2011). In order to obtain precise statistical values, the data was transformed into percentage values.

CHAPTER 4

RESULTS AND DISCUSSIONS

4.1 PERCENTAGE OF LIVE CUTTINGS

After 105 days in the greenhouse, the tungsten cuttings were removed from the substrate and data was collected. To analyse the live cuttings, the cuttings with green buds or leaves were counted and then separated by type of cutting (woody, semi-woody or herbaceous) and treatment (control, 50% and 100% AIB) (Photograph 1).

Photo 1- Tung cuttings removed from the substrate 105 days after transplanting
Source: the author.

The results representing the development of the pile according to the treatment received can be seen in Table 2.

ANOVA did not reveal a significant effect (P>0.05) for the live cuttings factor.

Table 2- Pile types for the live pile variable

Pile type	Averages (%)
Semi-woody	63.89 a
Woody	72.22 a
Herbaceous	77.78 a

Source: the author.
Averages followed by the same letter do not differ significantly.

The ANOVA revealed a significant effect (P<0.05) of the AIB concentration factor in relation to the live cuttings variable. At least one of the treatments differed significantly (P<0.05) from the others. The development of the cuttings subjected to the 100 per cent dose of AIB was superior to that of the cuttings that received 0 and 50 per cent of the dose (Table 3).

Table 3- Concentration of AIB for the live cuttings variable

Concentration of AIB	Averages (%)
Witness	47.22 b
50 AIB	66.67 b
100 AIB	100.00 a

Source: the author.
Averages followed by the same letter do not differ significantly.

As can be seen in photographs 2, 3 and 4, all the cuttings produced buds and leaves, but due to the high temperatures, these were eliminated and the cuttings died.

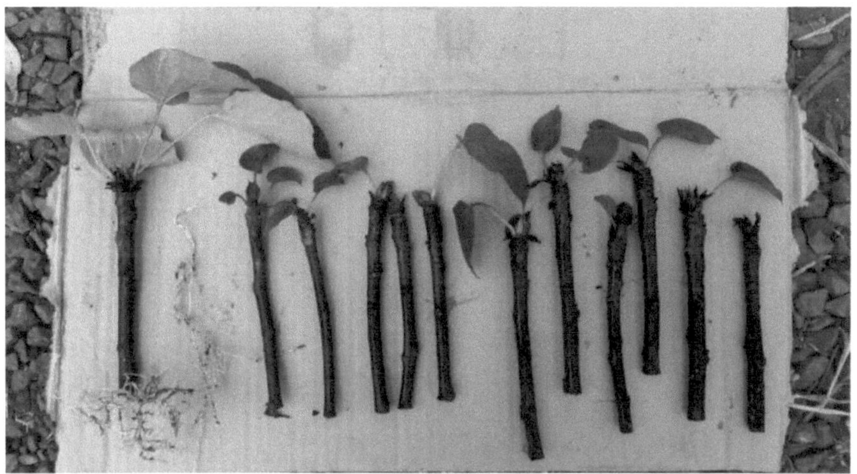
Photo 2- Herbaceous cuttings that received 100% of the AIB dose
Source: the author.

As with the herbaceous cuttings, the semi-woody cuttings also showed a drastic decrease in the percentage of live cuttings due to the rise in temperature inside the greenhouse in December, which reached a peak of 42°C.

Photograph 3- Semi-woody cuttings that received 100% of the AIB dose
Source: the author.

Photo 4 - Woody cuttings that received 100% of the AIB dose
Source: the author.

The herbaceous cuttings showed the highest percentage of live cuttings, despite the temperatures they were subjected to and the dose of AIB received.

According to Bastos et al. (2009), the use of AIB makes root formation more efficient, as it improves the quality of the roots formed by accelerating the rooting process, which leads to the production of better quality and uniform seedlings.

According to research carried out by Galvão (2000), AIB is the most effective growth regulator for root initiation via propagation by cuttings. Similarly, Lana et al. (2008) state that the use of different concentrations of AIB interferes significantly with the volume of green mass of the root system and the aerial part, as well as high dosages of growth regulators causing toxicity in the rooting of cuttings, and that the ideal dosage can vary according to the type of cutting used, species and cultivar.

According to Fachinello, Nachtigal and Hoffmann (2005) the number of live cuttings, as well as the percentage of callus, the length of the roots and the number of roots per cutting are factors that directly affect the yield of vegetative propagation.

4.2PERCENTAGE OF CUTTINGS WITH CALLUS

According to Hewitson and Barry (2017) root production begins as soon as the cuttings are separated from the mother plant, and the first reaction of the new plant is the formation of a tissue called callus. Although roots do not emerge from callus, it is a necessary element for the rooting of cuttings.

After identifying the cuttings that showed callus, the results were analysed by Analysis of Variance (Table 4), where the statistical analysis revealed no significant effect (P>0.05) for the factor types of cuttings in relation to the variable cuttings with callus.

Table 4 - Types of cuttings for the variable cuttings with callus

Pile types	**Averages (%)**
Woody	44.44 a
Semi-woody	52.78 a
Herbaceous	55.56 a

Source: the author.
Averages followed by the same letter do not differ significantly.

ANOVA revealed a significant effect (P<0.05) of the AIB concentration factor in relation to the cuttings with callus variable. At least one of the treatments differed significantly (P<0.05) from the others (Table 5).

Table 5 - Concentration of AIB for the variable cuttings with callus

Concentration of AIB	**Averages (%)**
Witnesses	30.56 b
50 AIB	41.67 b
100 AIB	80.56 a

Source: the author.
Averages followed by the same letter do not differ significantly.

The cuttings given 100% AIB had the highest number of cuttings with callus in the plots. The plots that received lower doses of AIB (0 and 50%) also showed these characteristics, but in smaller quantities (Photos 5 and 6).

Photograph 5- Woody cutting without AIB treatment with callus
Source: the author.

Photograph 6 - Semi-woody cuttings without AIB treatment with callus
Source: the author.

According to Davies (2017), the formation of callus on cuttings where growth hormones have been used is called indirect root formation, where callus is first produced to protect the wound made to the cutting and then adventitious roots are produced.

4.3 PERCENTAGE OF CUTTINGS WITH LEAVES

After counting the cuttings with leaves, the results were submitted to

Analysis of Variance (Table 6), where ANOVA revealed a significant effect (P<0.05) of the type of cuttings factor in relation to the variable cuttings with leaves. In other words, at least one of the treatments differed significantly (P<0.05) from the others. Herbaceous cuttings showed greater development of the aerial part compared to woody and semi-woody cuttings.

Table 6 - Types of cuttings for the variable cuttings with leaves

Pile types	Averages (%)
Woody	44.44 b
Semi-woody	66.67 b
Herbaceous	100.00 a

Source: the author.

Averages followed by the same letter do not differ significantly.

ANOVA revealed no significant effect (P>0.05) for the AIB concentration factor in relation to the variable cuttings with leaves (Table 7).

Table 7 - AIB concentration for the leafy cuttings variable

Concentration of AIB	Averages (%)
Witness	63.89 a
50 AIB	72.22 a
100 AIB	75.00 a

Source: the author.

Averages followed by the same letter do not differ significantly.

As can be seen in photographs 7, 8 and 9, all the herbaceous cuttings developed leaves on the aerial part regardless of the treatment received.

Photo 7 - Herbaceous cuttings without AIB treatment

Source: the author.

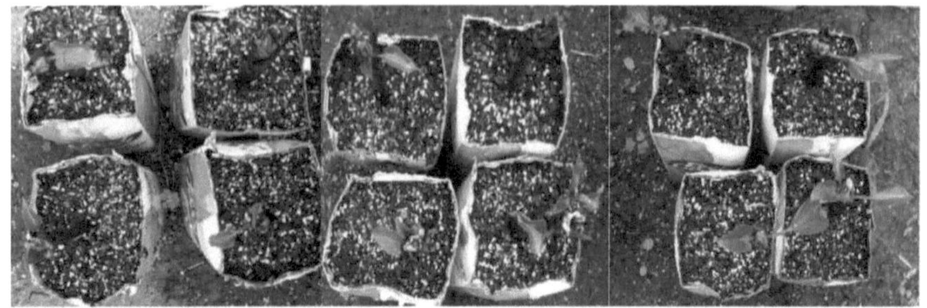
Photo 8 - Herbaceous cuttings treated with 50% AIB

Source: the author.

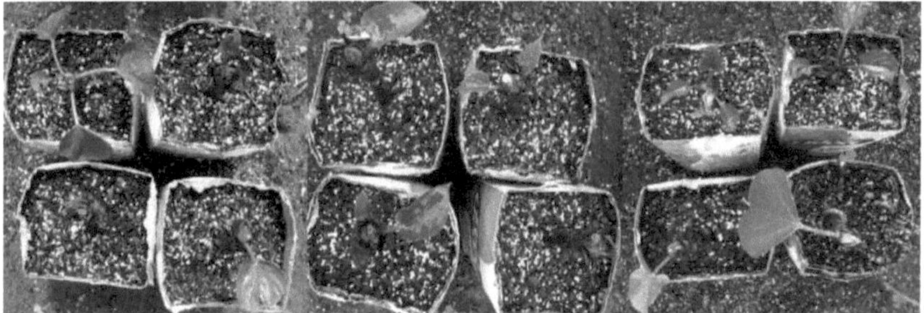
Photo 9 - Herbaceous cuttings treated with 100% AIB
Source: the author.

Research carried out by Dias, Ono and Duarte Filho (2011) shows that the presence of leaves on cuttings is one of the factors related to the rooting of cuttings. Photograph 10 shows a woody cutting treated with 100 per cent AIB with good development of the aerial part, which according to the same authors is a good indication of the development of cuttings.

Photo-array 10 - Woody cutting treated with 100% AIB
Source: the author.

According to Lima et al. (2007) the presence of leaves on the cuttings is important for rooting vigour, as they alter the availability of auxins and photoassimilates for root formation. In addition, the use of

auxins, combined with the presence of leaves, favours an increase in the percentage of rooted cuttings.

Like Chalfun et al. (2004) and Costa Júnior (2000), the presence of leaves influences the process of root formation, helping to transport rooting-promoting substances and promoting the loss of water through transpiration.

Studies carried out by Botin and Carvalho (2015) showed that the presence of leaf area on cuttings interferes with the rooting of cuttings, regardless of the thickness of the cutting, and that leaf emission guarantees survival and rooting.

4.4 PERCENTAGE OF ROOTED CUTTINGS

The percentage of rooted cuttings was 0.93%, i.e. only one cutting showed root development, this one being herbaceous and having been treated with 100% AIB.

During the time the cuttings were in the greenhouse, the plots were irrigated twice a day, the first at midday and the second in the late afternoon. The Tetra Pak boxes also had three holes drilled in the bottom to allow water to pass through freely.

It's important to note that the tetra pak boxes often had blocked holes and the cuttings from them began to show greater development of the aerial part. One of the herbaceous plants, which received a dose of 100% AIB and was exposed to greater water availability due to the obstruction of the holes in the boxes, showed fully developed roots (Photograph 11).

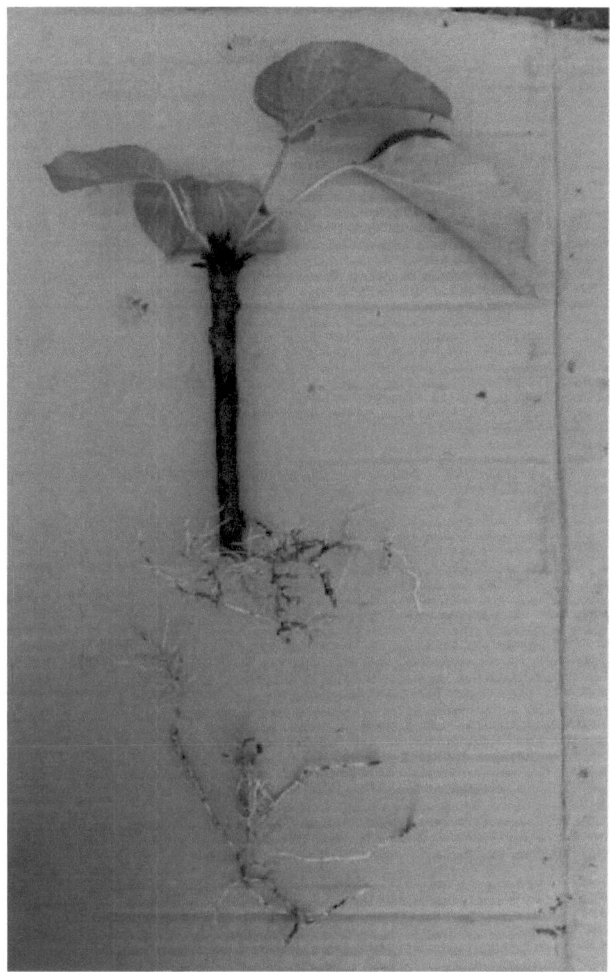
Photo 1- Herbaceous cutting treated with 100 per cent AIB
Source: the author.

According to Zeiger (2006), as well as influencing fruit production and development, auxins directly influence the formation of roots, those that sprout at the base of the stem, such as those seen in Photograph 11.

According to Fachinello, Nachtigal and Hoffmann (2005), the main factors affecting root formation are the type of cutting, the amount of hormone used or the hormonal balance of the plant, as well as the genetic rooting potential of the cultivar and the physiological condition of the parent plant. Lajús et al. (2007) state that the use of AIB in cuttings increases the rooting percentage and the quality of the root system of the cuttings.

According to Lajús et al. (2007), in order to obtain a quality seedling, the quality of the cutting's root system must be considered, as this factor will directly influence the plant's vigour and ability to survive.

CHAPTER 5

CONCLUSION

Indolebutyric acid contributes positively to the initial growth of tung (A/eurites *fordii* Hemsl) cuttings.

Indolebutyric acid does not significantly affect the percentage of cuttings with leaves at the concentrations used.

The cuttings treated with 100 per cent of the dose of indolbutyric acid showed the highest occurrence of callus.

The cuttings that received 100 per cent of the dose of indolbutyric acid had a greater number of developed roots than the cuttings that received 50 per cent of the dose and the control.

CHAPTER 6

FINAL CONSIDERATIONS

It is suggested that the cuttings be implanted at different times during the winter season, as well as transplanted during the spring, in order to avoid the burning of the shoots that occurs in December due to the high temperatures inside the greenhouse.

There is also a need to use temperature control mechanisms, such as shades or extractors, in the structure where the cuttings will be carried out.

There is a need for further studies to obtain information to prove the adaptation and development of tung in the far west of Santa Catarina. To this end, we suggest planting woody, semi-woody and herbaceous cuttings of tung in different locations in the far west of the state, especially in non-mechanised areas, with the aim of obtaining precise results regarding its adaptation as an alternative source of supplementary income for rural properties.

CHAPTER 7

REFERENCES

ALMEIDA, Ivan Rodrigues de; SILVA, Sérgio Delmar dos Anjos e; WREGE, Marcos Silveira. Agroclimatic zoning of the Tungue crop in southern Brazil. **Embrapa Clima Temperado,** Pelotas, v. 358, dec.2012.

ALVES, Élio Jos; LIMA, Bezerra Marcelo; SEREJO, Janay Almeida dos Santos; TRINDADE, Aldo Vilar. **Micropropagated seedlings.** Embrapa. 2015.
Available at:
<http://www.agencia.cnptia.embrapa.br/Agencia40/AG01/arvore/AG01_8_41 020068 O54.html>. Accessed on: 04 June 2017.

ÁVILA, Dante Trindade de. **Tungue** (A/eur/tes ***fordii)*** **cultivation in Rio Grande do Sul: population characterisation, propagation and agronomic performance.** 2010. Dissertation (Master of Science) Federal University of Pelotas. Postgraduate Programme in Family Agricultural Production Systems, Pelotas, 2010.

BARROS, Daniel Silva. **Tungue oil: technical specification.** Diadema. 2016.
Available at: <http://labmixquimica.com.br/wp-content/uploads/2016/07/062_b_informacoes_tecnicas_oleo_de_tungue.pdf>. Accessed on: 17 May 2017.

BASTOS, Débora Costa; SCARPARE FILHO, João Alexio; LIBARDI, Marília Neubern; PIO, Rafael. Estiolation, incision at the base of the cutting and use of indole butyric acid in the propagation of carambole by woody cuttings. **Agrotechnical Sciences.** Lavras, v. 33, n. 1, p. 313- 318, Jan-Feb, 2009.

BERTOCCHI, Dirlei Francisco. **Tungue as an alternative for income and environmental preservation.** 2011. Dissertation (Master of Science) Federal University of Pelotas. Postgraduate Programme in Seed Science and Technology, Pelotas, 2011.

BOTIN, Andréia Alves; CARVALHO, Andréa de. Growth regulators in the production of forest seedlings. **Revista de Ciências Agroambientais.** Alta Floresta, MT, v.13, n.1, p.83-96, 2015. Available at: <http://www.unemat.br/revistas/rcaa/docs/vol13-1/10_artigo_rcaa_v13n1 a2015.pdf.>. Accessed on: 27 Apr. 2017.

BUTTERFLY OIL. **Tungue oil.** Viamão. 2017. Available at: <https://www.oleodelinhacaetungue.com.br/index.php/exemplo-de-pagina/oleo-de- tungue.html>. Accessed on: 18 May 2017.

CAMPOS, Geraldo Maia. **Practical Statistics for Teachers and Graduate Students.**
USP, São Paulo, 2000. Available at: <http://www.forp.usp.br/restauradora/gmc/gmc_livro/gmc_livro_cap05.html>. Accessed on: 12 December 2015.

CASTRO, Paulo R. Castro; VIEIRA, Elvis Lima. **Applications of plant regulators in tropical agriculture.** Guaíba. Livraria e Editora Agropecuária, 2001.

CHALFUN, Rafael; PIO, Rafael; RAMOS, José; GONTIJO, Tiago, TOLEDO, Marcela. Presence of leaves and apical bud in the rooting of herbaceous fig tree cuttings from de-budding. Lavras. **Revista Brasileira Agrociência**. v. 10, n. 1, p. 51-54, jan-mar. 2004.

COSTA JÚNIOR, Walter Henrique da. **Rooting of guava cuttings:** influence of physiological and mesological factors. 2000. 66f. Dissertation (Master's Degree in Agronomy) Escola Superior "Luiz de Queiroz", Universidade de São Paulo, Piracicaba, 2000.

CUNHA, Catarina Monteiro Carvalho Mori; PAIVA, Haroldo Moreira; XAVIER, Aloisio; OTONI, Wagner campos. The role of mineral nutrition in the formation of adventitious roots in woody plants. **Pesquisa Florestal Brasileira:n°58.** Embrapa, Colombo, 2009.

DALL'ORTO, Luigi Tancredi Campo. **Auxins and types of cuttings in the rooting of Camelliasinensis.** 2011. Dissertation (Master of Science) - Phytotechnics, Luiz de Queiroz College of Agriculture, Piracicaba, 2011. Available at:<http://www.teses.usp.br/teses/disponiveis/11/11136/tde-21092011- 104432/en.php> Accessed on: 13 Oct. 2015.

DAVIES, Fred. **Principies of propagation by cuttings.**
Department of Horticultural Sciences. Texas A and M University and College Station. Texas. 2017. Available at: <www.uvm.edu/~mstarret/plantprop/chapter9.pps>. Accessed on: 06 June 2017.

DIAS, João Paulo Tadeu; ONO, Elizabeth Orika; DUARTE FILHO, Jaime. Rooting of cuttings from black mulberry root cuttings. **Revista Brasileira de**

Fruticultura. JaboticabaL 2011. vol. esp, Oct. 2011. Available at: <http://www.scielo.br/pdf/rbf/v33nspe1/a90v33nspe1.pdf>. Accessed on: 06 June 2017.

DUKE James A. **Handbook of energy crops.** Purdue University, USA, 1983. Available at: <http://www.hort.purdue.edu/newcrop/duke_energy/Aleurites_fordii.html>. Accessed on: 09 September 2015.

BRAZILIAN AGRICULTURAL RESEARCH COMPANY - EMBRAPA. **Propagation methods *for* tung** (Aleurites ***fordii).*** Research and development bulletin n°187. Pelotas, 2013.

BRAZILIAN AGRICULTURAL RESEARCH COMPANY - EMBRAPA. **Tung oil is an option for making biodiesel.** Embrapa Temperate Climate. 2008. Circular no. 624, year 14. Available at: <http://www.cpact.embrapa.br/comunicacao/linha/2008/linha624.pdf>. Accessed on: 04 June 2015.

SANTA CATARINA AGRICULTURAL RESEARCH AND RURAL EXTENSION COMPANY - EPAGRI. **Climatological Atlas of the State of Santa Catarina.**

Florianópolis, 2001. Available at: <httpciram.epagri.sc.gov.brindex.phpoption=com_content&view=article&id=708&Item id=484>. Accessed on: 08 September 2015.

FACHINELLO, José Carlos; NACHTIGAL, Jair Costa; HOFFMANN, Alexandre. **Propagation of fruit plants.** Brasília, DF: Embrapa Information Technology, 2005. 221 p. Available at: <http://www.ebah.com.br/content/ABAAAA_poAJ/propagacao-plantas-frutiferas>. Accessed on: 06 June 2017.

FERREIRA, Daniel Furtado. **SISVAR - System for analysing variance.** Version 5.6. Lavras-MG: UFLA, 2011.

FERREIRA, Fabricio Alves. **Plant Hormones.** São Paulo. 2013. Available at: <http://www.mundoeducacao.com/biologia/hormonios-vegetais.htm> Accessed on: 02 October 2015.

FLOSS, Elmar Luiz. **Physiology of cultivated plants.** 5 ed. Passo Fundo: UPF, 2011. 734 p.

GALSTON, Arthur W. **The green plant.** 1ed. São Paulo: Perspectiva, 1974.

118 p.

GALVÃO, Antonio Paulo Mendes. **Reforestation of rural properties for productive and environmental purposes: a guide for municipal and regional actions.**
Embrapa Forests. Brasília, DF, 2000. 351 p.

GENERAL IRON FITTINGS. **Tungue oil.** São Paulo. 2017. Available at: <https://ironfittings.com.br/produto/oleo-de-tungue/>. Accessed on: 18 May 2017.

GOLFETTO, Deisy Camila; ZAN, Renato Andre; BARBOSA, Nathália Vieira; BRONDANI, Filomena Maria Minetto; MENEGUETTI, Dionatas Ulises de Oliveira. Study and application of the Tungue kernel (A/eurites *fordii)* in the production of biodiesel. **I Científica da Faculdade de Educação e Meio Ambiente (Faema).Ribeirão Preto,** n.2, p.55-68nov-abr .2011. Disponívelem : <http://www.faema.edu.br/revistas/index.php/Revista-FAEMA/article/view/56>.
Accessed on: 09 September 2015.

GOMES, Cesar Bauer; SOARES, Vanessa Nogueira Soares; CAMPOS, Elaine Marisa Berton; STOCKER, Cristiane Mariluz; SILVA, Sérgio Delmar dos Anjos.
Evaluation of the nematicidal effect of castor and Tungue oils on second-stage juveniles *of Me/oido gyneethiopica.* **III Brazilian Congress on Castor Oil, Energy and Ricinochemistry.** Salvador, 2008.

GOOGLE MAPS. **Unoesc - Maravilha - Avenida Doutor Orlando Valério Zawadzki - Universitário, Maravilha - SC.** 2017. Available at: <https://www.google.com.br/maps/@-26.7646756,-53.1960237,375m/data=!3m1!1e3>. Accessed on: 04 June 2017.

GONTIJO, Tiago Chaltein Almeida. Rooting of different types of acerola cuttings using indolebutyric acid. **Revista Brasileira Fruticultura.**

Jaboticabal, v. 25, n. 2, p. 290-292, Aug. 2003. Available at: <http://www.academia.edu/7186554/Enraizamento_de_diferentes_tipos_de_estacas _de_aceroleira_utilizando_%C3%A1cido_indolbut%C3%ADrico>. Accessed on: 09 September 2015.

HEWITSON, John; BARRY, Meatyard. Root callusing in cuttings. **Science and plants forschoools.** University of Cambridge. United Kingdom. 2017.

Available at: <http://www.saps.org.uk/saps-associates/browse-q-and-a/500-what-causes-root- callusing-why-is-it-that-some-in-some-varieties-some-of-the-cuttings-root-readily- while-others-develop-a-large-amount-of-callus-which-prohibits-rooting>. Accessed on: 06jun. 2017.

IBGE. **Municipal Agricultural Production: Temporary and permanent crops.**
Brasilia, 2012. Available at:
<ftp://ftp.ibge.gov.br/Producao_Ag ricola/Producao_Agricola_Municipal_%5Banual%5 D/2012/pam2012.pdf>.
Accessed on: 08 September 2015.

KAUTZ, Jacqueline; ROSA, Caroline; D'OCA, Marcelo G. Montes; CLEMENTIN, Rosilene Maria. Extraction of tung oil (Aleuritis *fordii)* for biodiesel production. In: SIMPÓSIO ESTADUAL DE AGROENERGIA, 20. 2008, Porto Alegre.
Electronic proceedings... Pelotas, 2008. Available at: <http://www.alice.cnptia.embrapa.br/bitstream/doc/866426/1/012.pdf>. Accessed on: 08 Sep. 2015.

LAJUS, Cristiano Reschke; SOBRAL, Lúcia Salengue; BELOTII, Alencar; SAVARIS, Marco Andre; LAMPERT, Silvana; DOS SANTOS, Sandra Regina Furini; KUNST, Taciane. Indolebutyric Acid in the Rooting of Woody Fig Tree Cuttings (Ficus *carica* L.). **Brazilian Journal of Biosciences.** Porto Alegre, v. 5, supl. 2, p. 1107-1109, jul. 2007. Available at: <file:///C:/Users/Downloads/880-3048- 1-PB.pdf>. Accessed on: 06 June 2017.

LANA, Regina Maria Quintão; *LANA,* Ângela Maria Quintão; BARREIRA, Sybelle; MORAIS, Thais Rezende; FARIA, Marcos Vieira. Doses of indolbutyric acid on the rooting and growth of *eucalyptus cut*tings *(Eucalyptus urophy//a).***BioscienceJournal.** Uberlândia, v. 24, n. 3, p. 13-18, jul-sept. 2008.

LEMOS, Eurico Eduardo Pinto de. **Introduction to plant hormones.** Brasilia. Embrapa Recursos Genéticos e Biotecnologia, 2000. 180 p.

LIMA, Daniela Macedo; ALCANTARA, Giovana Bomfim; BIASI, Luis Antonio; FOGAÇA, Lucia Alves. Influence of leaf stipules and leaf number on the rooting of semi-woody cuttings of native yellow passion fruit. **Acta Scientiarum. Agronomy**, Maringá, v. 29, p. 671-676, 2007.

MALHEIROS, Euclides Braga; PANOSSO, Alan Rodrigo. **Experimental**

designs. Handout: UNESP, 2015.

MELO, Berilo de. **Fruticultura: Estaquia.** Federal University of Uberlândia: Institute of Agrarian Sciences. Uberlândia, 2003. Available at: <http://www.fruticultura.iciag.ufu.br/reproducao7.htm>. Accessed on: 25 September 2015.

MUSSER, Rosimardos Santos; COUCEIRO, E. M.; ALBUQUERQUE, M. H. Effects of naphthaleneacetic acid on the rooting of semi-woody acerola cuttings in a micro-sprinkler system. In: CONGRESSO BRASILEIRO DE FRUTICULTURA, 19.1987, Campinas. **Electronic Proceedings...** Campinas, 1987. Available at: <http://www.bdpa.cnptia.embrapa.br/consulta>. Accessed on: 09 September 2015.

NETO, Manoel. **Raw materials for biofuels.** Brasília. 2008. Available at: <http://materiaprimas.blogspot.com/2008/07/tungue-planta.html> Accessed on: 17 May 2017.

OLIVEIRA, Rejane B., GODOY, Silvana A.P., COSTA, Fernando B. **Toxic plants: knowledge and preservation of accidents.** Ed. Holos. Ribeirão Preto- SP, p. 64, 2003.

ONO, Elizabeth Orika; RODRIGUES, João Domingos. **Aspects of the physiology of rooting stem cuttings.** Jaboticabal: FUNEP, 1996. 83 p.

PRATI, Patrícia; MOURÃO FILHO, Francisco de Assis Alves; DIAS, Carlos Tadeu dos Santos; SCARPARE FILHO, João Alexio. Semi-woody cuttings: a fast and alternative method for producing 'tahiti' acid lime seedlings. **Scientia Agrícola**, Piracicaba, v. 56, n. 1, p. 1-8, 1999.

PRETTO, Alexandra. **Detoxification of Crambe and Tungue bran and evaluation of the nutritional response of Jundiá (Ramdhiaquelen).2O13**. 175p. Dissertation (Doctorate in zootechnics) - Federal University of Santa Maria, 2013. Available at: <http://cascavel. ufsm.br/tede/tde_busca/arquivo.php?codArquivo=4718>. Accessed on: 09 September 2015.

RAMO, Thiago Soethe. **Plant growth.** Campo Grande, 2012. Available at: <http://www.portaleducacao.com.br/biologia/artigos/19546/reguladores-de-crescimento-vegetal> Accessed on: 25 Sep. 2015.

REITZ, Raulino. **Flora Ilustrada Catarinense: Euphorbiaceae.** ed. Itajaí:

Empasc, 1988, 408 p.

SILVA, Paulo Henrique Muller da. **Seedling propagation systems for forest essences. Institute of Forestry Research and Studies (IPEF).** Piracicaba, 2005. Available at: <http://www.ipef.br/silvicultura/producaomudaspropagacao.asp>. Accessed on: 15 October 2015.

SILVA, Sérgio Delmar dos Anjos; ÁVILA, Dante Trindade; AIRES, Rogério Ferreira, ÁVILA, Thaís Trindade. **Agronomic Performance of Commercial Tungue Plantations in Rio Grande do Sul.** Embrapa Clima Temperado, Pelotas, v. 186, dec.2013.

SILVA, Sérgio Delmar dos Anjos; ÁVILA, Dante Trindade; AIRES, Rogério Ferreira, ÁVILA, Thaís Trindade. **Propagation methods *for* tung *(Aleurites fordii).*** Embrapa Clima Temperado, Pelotas, v. 187, dec.2013.

SILVEIRA, Carlos Augusto Posser; FERREIRA, Luís Henrique Gularte; SILVA, Sérgio Delmar dos Anjos; ÁVILA, Dante Trindade de. **Evaluation of the immediate effect of phosphate fertilisation and plant variability on tung fruit production.** Embrapa: 2010. Available at: <http://ainfo.cnptia.embrapa.br/digital/bitstream/item/45238/1/008.pdf> Accessed on: 16 May 2017.

TAIZ, Lincoln; ZEIGER, Eduardo. **Plant Physiology.** 5 ed. São Paulo: Artmed, 2013. 820 p.

TEIXEIRA, João Batista; MARBACH, Phellipe Arthur Santos. **Phytohormones.** Universa. Brasília: Catholic University of Brasília, 2000. p.232.

TOFANELLI, Mauro Brasil Dias; RODRIGUES, João Domingos; ONO, Elizabeth Orika. Method of application of indolbutyric acid in the cuttings of peach cultivars. **Agrotechnical Sciences,** Lavras. V.27, n.5, p. 1031-1037, Sept./Oct., 2003.

University of São Paulo - USP. **Introduction to Plant Biology.** São Paulo, 2002.

UCZAI, Pedro (Org). **Inevitable new world.** Chapecó: Pallotti, 2009. p. 27-34

VERNIER, Rafaela Maria; CARDOSO, Susette Barros. Influence of indolbutyric acid on the rooting of cuttings in fruit and ornamental species. **Revista Eletrónica de Educação e Ciência,** São Paulo, v.3, n. 2. p.11-16,

2013. Available at: <http://fira.edu.br/revista/vol3_num2_pag11.pdf>. Accessed on: 08 Sep. 2015.

WENDLING, Ivar. Vegetative propagation. **Embrapa Forests.** 1st University Student Week: Forests and the Environment. 2003.

XAVIER, Aloisio. **Chloral silviculture** I: principles and techniques of vegetative propagation. Viçosa: UFV, 2002. 64 p.

ZEIGER, Eduardo. **Plant hormones.** 2006. Available at: <http://www.todabiologia.com/botanica/hormonios_vegetais.htm>. Accessed on: 02 October 2015.

Printed by Books on Demand GmbH, Norderstedt / Germany